Alan McKirdy has written many popular books and book chapters on geology and related topics and has helped to promote the study of environmental geology in Scotland. His other books with Birlinn include *Set in Stone: The Geology and Landscapes of Scotland* and *Land of Mountain and Flood*, which was nominated for the Saltire Research Book of the Year prize. Before his retirement, he was Head of Knowledge and Information Management at Scottish Natural Heritage. Alan is now a freelance writer and has given many talks on Scottish geology and landscapes at book festivals and other events across the country.

Orkney and Shetland

LANDSCAPES IN STONE

Alan McKirdy

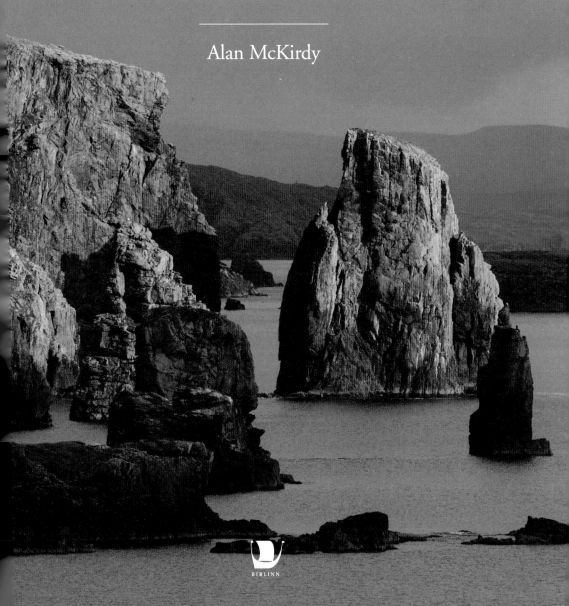

For John Gordon

First published in Great Britain in 2019 by
Birlinn Ltd
West Newington House
10 Newington Road
Edinburgh
EH9 1QS

www.birlinn.co.uk

ISBN: 978 1 78027 607 6

Copyright © Alan McKirdy 2019

The right of Alan McKirdy to be identified as the author of this work has been asserted by him in accordance with the Copyright, Designs and Patents Act, 1988

All rights reserved. No part of this publication may be reproduced, stored, or transmitted in any form, or by any means, electronic, mechanical or photocopying, recording or otherwise, without the express written permission of the publisher.

British Library Cataloguing-in-Publication Data
A catalogue record for this book is available on request from the British Library

Designed and typeset by Mark Blackadder

FRONTISPIECE.
Neap cliffs, Eshaness Peninsula, Shetland.

Printed and bound in Britain by Latimer Trend, Plymouth

Contents

	Introduction	7
	Orkney and Shetland through time	8
	Geological map	10
1.	Time and motion	11
2.	Early beginnings	13
3.	Colliding continents	14
4.	Death of an ocean	17
5.	Hot rocks: lavas, granites and gabbros	19
6.	Lake Orcadie and the Age of Fishes	21
7.	Mind the gap!	27
8.	The Ice Age and beyond	29
9.	Tsunami!	35
10.	Early occupation	36
11.	The machair – a living landscape	38
12.	Offshore oil and gas	39
13.	Places to visit	42
	Acknowledgements and picture credits	48

Introduction

The archipelagos of Orkney and Shetland are the most northerly outposts of the British Isles. Shetland represents the eroded roots of a mountain range that once soared to Himalayan heights. This fold mountain belt formed as continents collided some 420 million years ago and an ancient ocean closed. As a result, buckled and fractured rocks, for the most part, underlie the main islands of Shetland. These ancient rocks, the so-called basement on which later younger rock strata were subsequently piled, are pock-marked with granites and related rocks that were formed in the white heat of a cataclysmic continent-to-continent collision. Shetland forms part of a mountain range, known as the Caledonian Mountains, that extended northwards to Scandinavia and south-westwards through the Highlands of Scotland and onward to the Appalachians across the Atlantic Ocean. All these fragments were previously joined as one landmass.

A period of calm descended after these apocalyptic events. The lower slopes of the newly formed mountain range were partially drowned under a huge freshwater lake which extended over much of Orkney, into parts of Shetland and the north of Scotland. It was fed by rivers and streams that flowed from the surrounding high ground. There was no vegetation to bind the soils, so erosion was fast. Layer upon layer of sands, silts and muds built up in the lake itself, and sand dunes were created by fierce desert winds at the lake margins. This freshwater body, known as Lake Orcadie, existed for almost 10 million years and was home to a wide range of primitive fish species.

Hundreds of millions of years elapsed between the disappearance of Lake Orcadie and the next big event to leave its mark: the Ice Age. Great fingers of ice and snow extended south from the North Pole and the area was completely submerged beneath a blanket of ice.

In this book, we explore these stories in more detail and explain how the landscapes of Orkney and Shetland came to be the way we recognise them today.

Opposite.
The Old Man of Hoy.

Orkney and Shetland through time

Period of geological time	Millions of years ago	Scotland's global position	Environments and events in Orkney and Shetland
Anthropocene	Last 10,000 years	57° N	This is the period when our species *Homo sapiens* first lived in the area. Earliest evidence of human occupation at Skara Brae on Orkney dates back some 5,000 years. A tsunami struck 8,100 years ago and temporarily overwhelmed the islands and their ecosystems.
Quaternary	Started 2 million years ago	Present position of 57° N	• **11,500 years onwards** – the ice melted as the climate started to warm due to climate change. • **12,500 to 11,500 years ago** – the climate became very cold, and tundra-like conditions prevailed. • **14,700 to 12,500 years ago** – for a brief interlude, temperatures became warmer. • **18,000 years ago** – Shetland was covered by ice, but Orkney was largely ice-free. • **Before 29,000 years ago** and for a period approaching the last 2.7 million years, there were prolonged periods when thick ice sheets covered the islands. These advances of the ice were punctuated by warmer interludes, known as inter-glacials, when the temperatures rose to levels similar to those of today.
Neogene	2–24	55° N	Conditions were warm and temperate during these times, but temperatures fell as the Ice Age approached.
Palaeogene	24–65	50° N	Between 65 and 60 million years ago, the ancient continent of Pangaea was split asunder by plate movements. Eight major volcanoes were active during these times, but there is no evidence of these events here.
Cretaceous	65–142	40° N	Warm shallow seas covered the area, but no rocks of this age are preserved here.

Period of geological time	Millions of years ago	Scotland's global position	Environments and events in Orkney and Shetland
Jurassic	142–205	35° N	The rocks that formed the basis for the oil and gas industry were deposited in the North Sea and to the west of Shetland. But no deposits of this age are preserved on land.
Triassic	205–248	30° N	Desert conditions prevailed across Scotland, but no rocks of this age are preserved here.
Permian	248–290	20° N	Desert conditions were widespread, but no rocks of this age are preserved here.
Carboniferous	290–354	On the Equator	'Scotland' was located at the Equator at this time, but no rocks of this age are preserved here.
Devonian	354–417	10° S	Lake Orcadie was established at this time and laid the foundations for Orkney. Thick sandstone deposits are also found on Shetland, accompanied by extensive volcanic activity.
Silurian	417–443	15° S	Large upheavals created the Highlands of Scotland.
Ordovician	443–495	20° S	'Scotland' was located on the northern shores of a long-disappeared ocean.
Cambrian	495–545	30° S	No deposits of this age are found here.
Proterozoic	545–2,500	Close to South Pole	The Moine and Dalradian rocks were formed during this period.
Archaean	Prior to 2,500	Unknown	The Lewisian gneisses date from these times. The age of the Earth is around 4,540 million years.

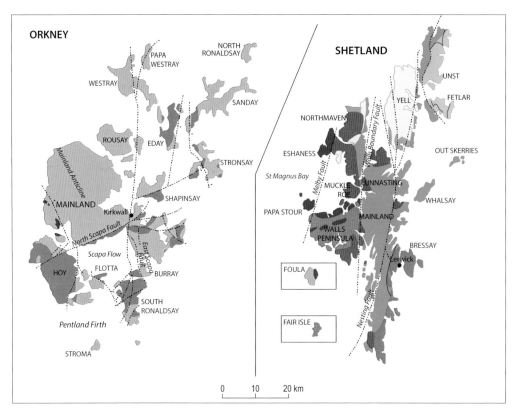

Key

SEDIMENTARY AND VOLCANIC ROCKS OF THE OLD RED SANDSTONE
- Orkney Sandstone Group with volcanic rocks and Bressay Formation — Upper Devonian
- Caithness Flagstone Group and Melby Formation — Middle Devonian
- Walls Formation and Sandness Formation with volcanic rocks shown in blue — Lower Devonian

INTRUSIVE IGNEOUS ROCKS (of Silurian to Devonian age)
- Granite and related rocks
- Mantle and ocean-floor rocks

METAMORPHIC ROCKS (all of Precambrian age)
- Dalradian
- Moine
- Age uncertain
- Lewisian
- Basement rocks of Orkney

- ········ FAULT
- ─── THRUST

The geological map of Shetland represents a complex coming together of rocks from different eras. The elements of this bedrock patchwork are separated one from another by a series of north/south-trending faults or cracks in the Earth's outer layer (crust). The most ancient rocks of Shetland are among the oldest in the world, broadly equivalent to the Lewisian gneisses of the Western Isles. Granites, coloured red on the map, punched their way through this ancient basement of rock. Unst and Fetlar are a magnet for visiting geologists: the rocks of these islands are interpreted as the ancient floor of an ocean that existed for 250 million years, but is no more. **The isles of Orkney**, in contrast, have a largely uniform bedrock. These islands are almost entirely built from sandstones, laid down in a long-disappeared freshwater lake that occupied much of northern Scotland and also stretched northwards to cover the south-eastern part of Mainland Shetland. It's an amazing geological story.

1
Time and motion

Before the geology of these far-flung outposts of Scotland can be tackled in a meaningful way, we need a perspective on the immensity of geological time and an appreciation of the dynamic nature of the way our planet's surface operates. Without that understanding, it is difficult to make sense of the geological story that follows.

Time

We tend to see past events through the prism of our recorded historical record, monuments and artefacts. For many people, Stone Age people or the Roman occupation seem to be the furthest back we can extend that record. Understanding geological events is another challenge entirely. Millions, even billions, of years before the present day, is the standard timescale for the geologist. Working out the date of these far-off events, for example the age of a fossil, or a significant event such as the eruption of a lava flow, isn't the stuff of scientific fantasy. We can now achieve an accuracy of plus or minus only a few million years. The business of dating rocks and events was pioneered in Edinburgh by an exceptional scientist: Professor Arthur Holmes. He hailed originally from Gateshead, but he did his finest work while he was Regius Professor at Edinburgh University in the 1950s and 1960s.

His work, along with others who came after, allowed geological time to be subdivided into manageable chunks, known as periods. By international agreement, each geological period is date-stamped with a beginning and end date in millions of years. This allows rocks and geological events to be placed in date order. As a result, the sequence of events that created the landscapes we see around us can be established with a high degree of reliability. In the table on pages 8 and 9 of this book, the time sequence is arranged in this way and can be seen as the pageant of geological events that led to the formation of Orkney and Shetland today. Geologists start with the oldest events and work through geological time to more recent events. So, in this book, we are on a journey through geological time from first to last.

Motion

We are also on a second journey: this time across the globe. Scotland's journey started near the South Pole around 2,000 million years ago and it has been moving northwards ever since. The driving force for this movement is located deep below our feet – the Earth's core. Heat leaks from this super-hot ball of iron and nickel at the centre of the planet and sets up a convection motion in the overlying layer known as the mantle. The ground beneath our feet is literally dragged along as a result. The Earth's surface, or crust, is divided into seven large chunks and many smaller ones, known as tectonic plates. They move independently of each other at an imperceptibly slow rate. On average, this is about 6cm per year, but over many millions of years, this can move continents from one side of the globe to another. The land that became Scotland has moved through every climatic zone on the planet, from the deep freeze of the polar regions to the scorching heat of the Equator.

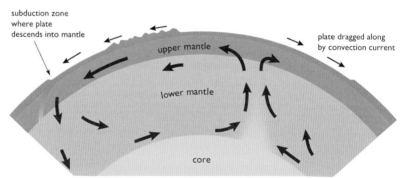

Heat radiates from the Earth's core creating convection currents in the overlying mantle. This driving force moves the overlying continents around the globe, so the geography of the world has been changing continuously throughout the history of the Earth.

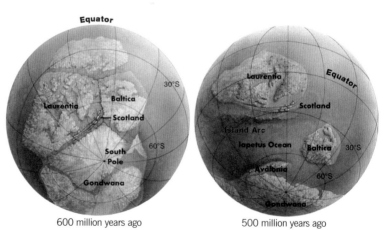

These two globes illustrate the constantly changing geography of the early planet. Around 600 million years ago, the continents were all clustered around the South Pole, but later they split as a new ocean became established – the Iapetus Ocean. This is shown in the second globe, which illustrates the geography of the world 500 million years ago. 'Scotland' and 'England' were located on opposite sides of the ancient ocean.

600 million years ago   500 million years ago

2
Early beginnings

The oldest rocks of Shetland are similar in many respects to the Highlands and Islands of Scotland. Isolated patches of rocks that are comparable to the Lewisian gneisses of the Outer Hebrides are located on Mainland Shetland (coloured pink on the geological map). These ancient rocks date back almost 3,000 million years into Shetland's dim and distant geological past. They were significantly altered, folded and faulted deep in the Earth's crust where they were subjected to high temperatures and pressures – a process known as metamorphism. Some of these rocks were originally ancient muds, limestones and volcanic lavas, but have been much changed over the last few billion years to metamorphic rocks, known as gneisses (pronounced 'nices').

They are found in close association with Moine rocks (coloured yellow on the geological map). The Moine rocks were laid down under water as a thick series of sands and muds that were also later buried at depth in the Earth's crust and significantly changed as a result. These events took place around 950 million years ago. The island of Yell is the best place to find rocks of this age. These Moine schists (a type of metamorphic rock) have been described by geologists as 'monotonous'. They are uniform in colour, entirely without fossils and offer little in the way of variety. But, nonetheless, they are an important part of the geological story.

Isle of Yell, Shetland.

3
Colliding continents

The bedrock of Shetland is a remnant of an impressive mountain chain. It is the product of a collision of continents that took place over many millions of years and ended around 420 million years ago. What remains of these Caledonian Mountains stretches over two continents – from the west coast of Norway, through Shetland into the Highlands of Scotland and across the Atlantic Ocean to the Appalachians. Shetland occupies a pivotal position in linking the geology of these now widely distributed places.

The rocks that made up the Caledonian Mountains are known as Dalradian schists and form the largest extent of altered rocks in Shetland. They occupy much of the central area of Mainland, Whalsay and Out Skerries. They were originally a mixed series of sand, muds, a few limestones, together with volcanic lavas and associated ashes, that accumulated on the floor of a major and now long-disappeared

The Caledonian Mountains would have rivalled the present-day Himalayas in size and scale when they were first formed. The land that became Shetland sat almost midway along this mountain range and has played a key role in understanding the processes at work in creating this natural wonder. Later movement of the continental plates split this mountain range asunder, and the component parts of that larger whole now sit on opposite sides of the Atlantic Ocean.

As continents collide, the sediments that accumulated on the ocean floor are folded and buckled to form a mountain chain. The older rocks of Shetland were formed in this way and were as impressive as the Himalayas are today when they were first formed around 420 million years ago.

ocean – the Iapetus Ocean. Around 600 million years ago the continents of the world were grouped around the South Pole. This landmass subsequently splintered into three separate areas of dry land, and proto-Scotland was located on the southern shore of the continent that geologists have named Laurentia. This continent also included an early rendering of North America and Greenland. The land that was to become England and Wales lay on the opposite side of an ever-widening ocean.

This was the arrangement of the continents around 500 million years ago. Sometime later the Iapetus Ocean started to close, bringing Laurentia (and 'Scotland') closer to 'England and Wales'.

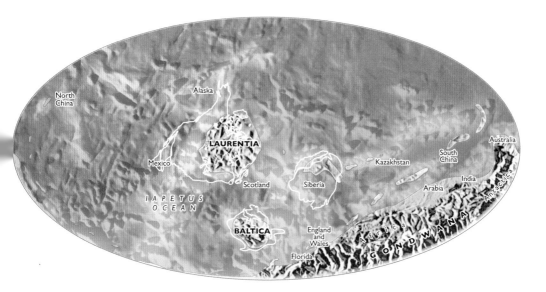

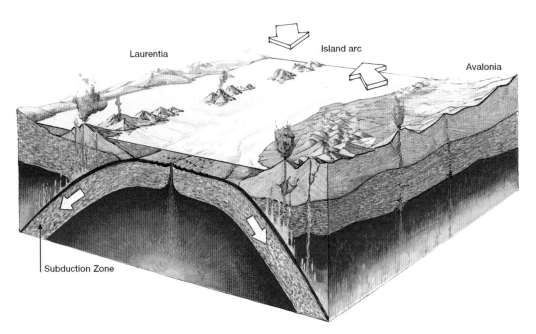

The leading edge of one of the tectonic plates dipped below the landmass that contained 'Scotland', causing the Iapetus Ocean to close. This process is known as subduction. As this leading edge descended further into the Earth's mantle, the rocks overlying it melted and bubbled back to the surface to form a chain of volcanoes known as an island arc. As the two landmasses approached each other, the layers of sands, muds, limestones and lavas that had accumulated on the ocean floor over a 250-million-year period were squeezed as if in the jaws of a vice. A fold mountain belt, similar in style to many across the globe (such as the Himalayas, Alps and Urals), was formed as a result.

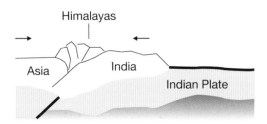

India was driven northwards and collided with Asia around 55 million years ago. The layers of sediments that had accumulated in the intervening ocean floor were thrust upwards forming the spectacular mountain chain we see today. The Caledonian Mountains, including Shetland, Highland Scotland, parts of Norway, Greenland and the Appalachians, were created in a similar manner.

4
Death of an ocean

As the Iapetus Ocean started to close about 500 million years ago, a little part of the ocean floor and a fragment of the underlying Earth's mantle became detached and was thrust upwards towards the surface. Ordinarily these rocks would have been subducted with the rest of the

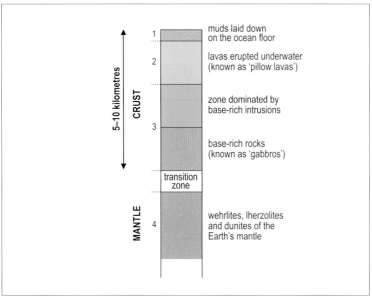

The ocean floor, today and in the geological past, has a consistent composition. This diagram shows a slice through the rocks that floor the world's oceans today, but would have also made up the ocean floor 500 million years ago. The ocean crust, consisting of calcium and magnesium rich rocks (mainly basalt and gabbro) has a variable thickness of between 5 and 10 kilometres. Below these layers lies the Earth's mantle. Separating the crust and the mantle is a transition zone known as the Mohorovičić discontinuity or more affectionately as the 'Moho'. It is named after the Croatian scientist of that name who discovered it. Very dense rocks of the mantle lie beneath the Moho with picturesque names such as harzburgite, wehrlite, lherzolite and dunite. What is so special about Unst and Fetlar is the exposure at the Earth's surface of the Moho and the underlying dense black rocks of the mantle. Nowhere else in Britain can you walk across such extensive exposure of the lower reaches of the Earth's crust and mantle. That experience was made possible by the death of the Iapetus Ocean, a process that started 500 million years ago.

17

There is a unique flora associated with the Unst ophiolite complex. Many of the plants that grow here are nationally rare. The soils are deficient in phosphorous, nitrogen and potassium and carry higher concentrations of heavy metals such as chromium and nickel. The local flora that thrives here is clearly tolerant of these chemical peculiarities. Star attraction in this botanical diversity is Edmondston's chickweed (first recorded in 1837 by local botanist Thomas Edmondston). Unst is the only place in the world where this plant grows.

ocean floor as the ocean narrowed and continents collided, but somehow they were held up (or obducted as geologists describe the process) and remained at a higher level in the crust. So we can now see and walk across a representative sliver of rocks that once lay beneath the Earth's crust. The deepest boreholes can't access rocks below the Earth's crust, so we only know how the Earth is structured by remote sensing. This rarest of phenomena, a fragment of oceanic crust and underlying mantle, is known as an ophiolite complex. Ophiolite rocks have been intensively studied since their significance was first established. There are good examples on the eastern side of Unst and Fetlar, which geologists regard as one of the finest and most accessible examples of ophiolite rocks in Europe.

Around 420 million years ago, the bedrock was subjected to more torment. Three major faults or tears in the Earth's crust, and a few minor ones, were initiated. Their north-north-east orientation and location is shown on the geological map on page 10. The horizontal movement along these faults is estimated to have been around 170km, so the component parts of the jigsaw that makes up the Shetland bedrock were considerably rearranged at that point. Later there was further reactivation of these faults at the end of the Devonian Period and again around 200 million years ago. Today, in the UK, where we are mercifully free from destructive earthquakes, we are shocked if earth tremors make our windows rattle. But when this land was being assembled 400 million years ago, seismic shocks were massively more severe and frequent than they are today.

5
Hot rocks: lavas, granites and gabbros

In Shetland, the final chapter of these turbulent events was the eruption of a thick sequence of volcanic lavas and the injection of a series of granites that punched from deep within the Earth's mantle into the upper reaches of the Earth's crust (geologists call these intrusions). The lavas and granites associated with these turbulent times are seen around St Magnus Bay and Northmaven in western Shetland.

Along the dramatically cliffed Eshaness coast of western Mainland Shetland, thick accumulations of lava flows, piled one on top of another, make for a spectacular sight. The lava sequence provides evidence of explosive volcanic activity. These rocks were erupted as thick clouds of red-hot ash and volcanic debris similar to those that smothered Pompeii almost 2,000 years ago. Shards of volcanic glass, pumice and thick layers of red-hot ash were erupted to form a pile around 500m thick. The landscape over which these ash clouds were erupted was largely devoid of life, so these deadly volcanic episodes created no casualties.

The cliffs at Eshaness, western Shetland.

These events were partly related to the closure of the Iapetus Ocean and are dated as having taken place well into Devonian times. The intrusion of the granites and the eruption of the lavas took place over many millions of years, in fact an astonishing 40 million years or so. The composition of the molten rock or magma predominantly gave rise to granites, but some gabbros (dark-coloured rocks rich in calcium and magnesium) are also present.

Granites underlie Ronas Hill, the highest point in Shetland, which rises from Ronas Voe in west Shetland to a height of 450m. Other intrusions include the Sandsting granite that occupies the southern part of the Walls Peninsula. This was injected as a series of molten granite sheets that metamorphosed or 'cooked' the sedimentary layers into which they were introduced. The Skaw Granite in north-eastern Unst is intensively fractured, so the interpretation here is that it was transported to its present position along a fault line. There are other granites across Mainland Shetland including the pink and green granites of Hildasay that were quarried and shipped to Australia as ballast and subsequently used as building stone. These granites don't form significant landscape features. Soils overlying these rocks are thin with limited fertility.

In Orkney, volcanic activity of this time, around 380 million years ago, was much more limited. On the east side of Mainland Orkney and also on the nearby island of Shapinsay, thin lava flows have been identified, indicating the presence of active volcanoes. Also, the iconic Old Man of Hoy sea stack is supported on a plinth of lava, perhaps giving this transient structure a bit more stability than if the whole edifice had been built from sandstone.

Ronas Hill in the distance is the highest point in Shetland.

6
Lake Orcadie and the Age of Fishes

After the mayhem of the continental collisions that created the Caledonian Mountains came a period of relative calm. The new arrangement of land and sea was markedly different from before. The separation between 'Scotland' and 'England' was no more and a larger continent consisting of what would become the British Isles, North America and Greenland was created. Geologists have called this new landmass the 'Old Red Sandstone continent' or the more scientifically correct name of Laurussia.

The new Caledonian Mountains soared to heights comparable with any uplands areas on the planet today. The Old Red Sandstone continent was initially a very unstable land. Earthquakes regularly shook the continent. Erosion by wind and water was rapid, and the tops of the mountains were quickly planed down. The evolution of plants was at a very early stage during Devonian times, so there was no vegetation to bind the steep rock faces and mountain slopes. Boulders, cobbles, sands and muds tumbled down the slopes, and fast-

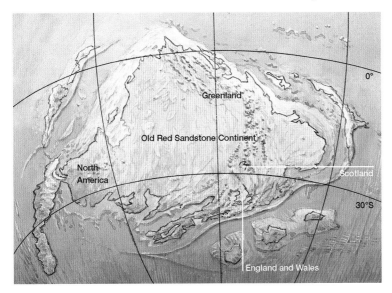

The closure of the Iapetus Ocean gave rise to a new and transient landmass that geologists have called the Old Red Sandstone continent. The new mountain belt can be seen running through the outline of Scotland and northwards to Greenland.

This was the scene some 400 million years ago when the Old Red Sandstone continent was created. The climate was very arid, consistent with the continent's position close to the Equator.

This illustration shows the extent of Lake Orcadie – from the fringes of what became the Moray Firth coast, extending northwards to occupy the area we now recognise as Orkney and Shetland. It also extended eastwards and had occasional connections to the sea beyond the Old Red Sandstone continent. The present day coastline is shown as a dotted line for reference, but the distribution of land and water was very different 400 million years ago.

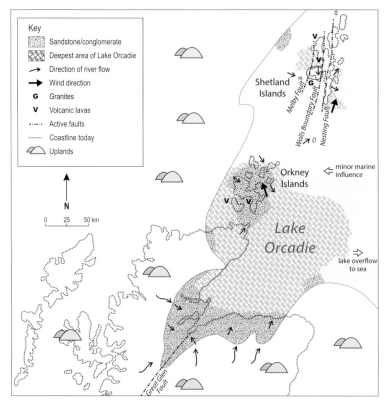

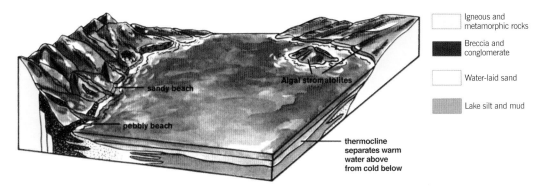

Environments and deposits in Lake Orcadie during the Middle Devonian. Lake Orcadie existed for about 10 million years. In that time, around 5km of sandstones, muds and associated sediments were laid down. The lake dried out from time to time, but scientific study suggests that the lake was around 80m at its deepest point.

flowing rivers transported this material to lower ground. It accumulated as layers of sandstone, mudstone and conglomerate that comprise the deposits we now call the Old Red Sandstone.

A huge freshwater lake subsequently developed between the towering mountains of the new continent. This was a sump into which the rivers and streams flowed. It extended eastwards as far as the present-day Norwegian coastline and had brief links with the sea from its south-eastern margin. The lake was never particularly deep at any time during its 10-million-year existence, probably around 80m at the deepest point, but in area it covered an estimated 50,000 square kilometres. That is difficult to imagine, but the diagram on page 22 should help to define its extent.

The lake sediments give a clear indication of the prevailing environmental conditions at any given time and in any particular place. By aggregating the information from careful study of various localities across the area, a picture emerges of the changing environmental scene in Lake Orcadie.

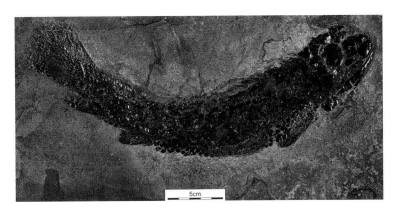

The remains of many different species of fish have been recovered from the lake sediments of Lake Orcadie. The scientific name of this specimen is *Gyroptychius agassizi*.

This specimen of the lobefin fish *Osteolepis macrolepidotus* was recovered from Cruaday Quarry on Mainland Orkney.

Deep water is indicated by the occurrence of very fine sediments. Many layers contain the remains of past life, predominantly fish that sank to the bottom of the lake after they had died. These very finely banded sedimentary layers indicate a very slow rate of sediment accumulation on the floor of the lake, probably associated with rapid evaporation from the lake itself. The fossil fish found here are quite remarkably well preserved as a result of the development of oxygen-poor conditions in the deeper parts of the lake. This inhibited scavengers, so the fish remained intact and undisturbed.

The fossil fish recovered from the Old Red Sandstone of Orkney and Shetland are among the most diverse and varied of this age to be

Pterichthyodes milleri – this species of fish links sites across Lake Orcadie as being of the same age.

found anywhere in the world. They were also some of the earliest fossil specimens to be scientifically described and classified.

Cruaday Quarry, on Mainland Orkney, has been worked for its fossil fish since the 1830s. Professor Traill from Edinburgh University was the first to publish his findings on these fossils. Since that time, some 15 species of fossil fish have been described from this place. The science is now well established, and each specimen is classified thus: for example **Class** (*Osteichthyes* – bony fish), **Genus** (*Osteolepis*) and **Species** (*macrolepidotus*).

Such rigour in naming and classifying fossil finds allows specimens to be compared and contrasted across the globe, so scientists can establish how populations of the same animals developed in different parts of the world. Melby fish bed in west Shetland illustrates this point very neatly. It contains a diagnostic fossil fish *Pterichthyodes milleri* that is also found in Caithness and Orkney, so the fish beds from these various locations can be linked together as being of the same age.

Where the lake had dried out, desiccation cracks developed in the layers of mud. These cracks only develop when the layers of mud have been baked by the sun. So when desiccation cracks are found in ancient sediment such as these, we can say for certain that they were exposed to the sun and not submerged by the waters of the lake. Information such as this is important in plotting the changing shoreline of Lake Orcadie throughout its 10-million-year history. These events played out around 380 million years ago during Middle Devonian times.

Closer to the shore of the lake, there is a greater predominance of sandy layers. When the lake dried out completely, desert conditions were established across the area. The Old Red Sandstone continent

These geometric shapes form when layers of mud are dried in the sun. Where these structures are found in the geological record of Orkney and Shetland, they are taken as an indication that mud deposits were temporarily exposed to the air when the area covered by Lake Orcadie shrank. These mud cracks have helped geologists to map the changing extent of Lake Orcadie through time.

This spectacular sea stack at Castle of Yesnaby, Orkney, shows, in its upper reaches, structures that are known as cross-bedding. These angled layers are slices through ancient sand dunes that fringed Lake Orcadie. The sand dunes of the Sahara desert would be a good model for what these dunes would have looked like 380 million years ago. It is much more recent geological activity that has caused these layers now to form part of a sea stack.

lay 10° south of the Equator at the time, so such conditions are to be expected.

The halfway point between Orkney and Shetland is marked by the picturesque Fair Isle. It is almost entirely made from rocks of the Old Red Sandstone. The layers vary in character from conglomerates made from pebbles and cobbles to sandstones and mudstones deposited in the deeper waters of Lake Orcadie. The rock strata were later sliced by a series of six faults that run west-north-west across the island. These weaknesses were exploited later by pulses of molten rock that flowed from deep within the Earth, and later still by the sea. Ferocious storms regularly lash this place, and the coastline has been carved into a variety of spectacular landforms such as high cliffs, geos (long narrow clefts in sea cliffs), stacks and skerries (rocky islands).

Fair Isle is also built from pebbles, cobbles, sand and mud laid down in Lake Orcadie.

7
Mind the gap!

A yawning gap of geological time stretched from the time of Lake Orcadie to the next event to leave its mark on Orkney and Shetland: around 400 million years in fact. During that period, the land that became Scotland moved from a position south of the Equator and journeyed northwards. In so doing, it moved through many of the Earth's climatic zones. While it remained near the Equator during Carboniferous times, tropical rainforests clothed much of the Scottish landscape; the remains of these forests gave rise to the coal that powered the Industrial Revolution many millions of years later. Desert conditions came next in Permian and Triassic times.

Then the sea level rose dramatically and dinosaurs roamed across Skye – Scotland's very own Jurassic Park. During Cretaceous times, sea levels rose further and all but the highest ground disappeared under the ocean. Later still, during a major rearrangement of global

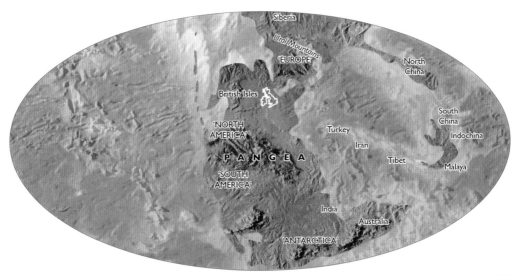

By 250 million years ago, 'Scotland' had continued its northwards progress and sat 30° N of the Equator. It formed part of a landmass that stretched from pole to pole called Pangaea – 'All Earth'.

Tropical rainforests, deserts, inundation by sea and more recent volcanic activity than the Devonian times all left their mark elsewhere in Scotland, but not on Orkney and Shetland.

geography, the North Atlantic Ocean was formed and a series of volcanoes roared into life. We recognise them today as the eroded remnants of the formerly active volcanoes of Rockall, St Kilda, Skye, Rum, Ardnamurchan, Mull, Arran and Ailsa Craig. Astonishingly there is no evidence of any of these events in Orkney or Shetland! Evidence for these long-lost environments may have at one time been here, but events that were about to happen (described next) may have removed all trace. Around 2.7 million years ago, the Ice Age was about to begin.

8
The Ice Age and beyond

As the Ice Age approached, the climate cooled dramatically and icy tentacles extended south from the North Pole. Soon, much of what would become northern Europe was covered in a blanket of ice and snow that was several kilometres thick. The ice sheets waxed and waned over the two-million-year period that the Ice Age lasted. Each cold period was separated from the next by an inter-glacial period where temperatures were warmer, much as they are today. The ice sheets and associated glaciers had a dramatic effect on the landscapes that we see today. Uplands were planed down to a lower level, and deep channels were cut between the islands.

As well as eroding the landscape and gouging great trenches between the islands, the ice also transported boulders, rocks and sand huge distances and then dumped them when the ice melted. One of the most famous of these boulders, the so-called 'glacial erratics', is the Dalsetter Stone, found near Sumburgh, Shetland.

The particular variety of granite that the Dalsetter Stone is made from can be matched to the bedrock near Oslo in Norway. There are

The two-tonne Dalsetter Stone, deposited in Shetland by an ice sheet, was too heavy to move so it was incorporated into a drystane dyke.

This was the scene 18,000 years ago. Orkney was entirely ice-free but Shetland retained a small ice cap. The Scandinavian ice sheet was still a major influence, but it had retreated significantly since the time it deposited the Dalsetter Stone.

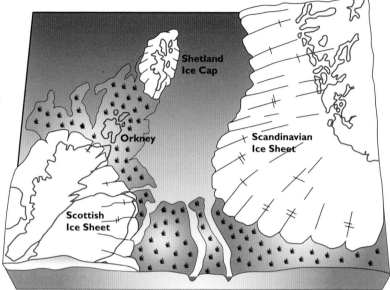

no possible local sources of this stone in Shetland, so we can say with certainty that the Scandinavian ice sheet must have stretched across the North Sea, carrying with it chunks of rock carved from the underlying Scandinavian bedrock until the climate warmed up and the ice melted.

The glaciers on Shetland laid down a thin veneer of glacial debris, mainly sands and gravels. Buried beneath these are layers of peat that contain fragments of pine trees and pollen grains. These organic layers have been dated to around 380,000 years ago when the climate was substantially warmer than full-on glacial conditions. So Shetland would have had a substantial woodland cover at that time. Peat layers found near Sel Ayre on the Walls peninsula, Mainland Shetland, contain pollen grains that indicate a heathland habitat. At Garths Voe on Shetland, heathland habitat was replaced by the development of willow, birch and hazel between 8,000 and 5,000 years ago. So the changing environments and habitats during and since the Ice Age can be accurately modelled by a careful study of pollen grains and other organic remains recovered from the layers of peat. Study of these organic layers has added considerably to our understanding of environmental change in these more recent times.

The blockfield, an area strewn with rocks and boulders, covers the slopes of Ronas Hill on Mainland Shetland. It tells of extreme environments that developed after the ice finally melted around 11,500 years ago. The climate remained sub-arctic, and this blockfield and the stone

THE ICE AGE AND BEYOND

The blockfield formed on top of Ronas Hill and the stone stripes on Ward Hill (in the distance) both formed in response to the extreme conditions that prevailed from the end of the Ice Age to the present day.

stripes on Ward Hill in the far distance (illustrated in the photograph above) are indicative of such harsh environments. Similar features and habitats are found today in sub-arctic areas of northern Norway.

A drowned coastline

Sea levels rose worldwide as the glaciers melted at the end of the Ice Age. Much of Orkney and Shetland's coastal area was drowned under the rising seas. The heavily glaciated landscapes were flooded as the sea levels climbed higher. As a result, great inlets were created that cut to the heart of Mainland Shetland particularly. We recognise these

The valleys, originally cut by rivers and streams and later considerably enhanced by the gouging action of the ice, were flooded as sea levels rose at the end of the Ice Age. Thin ribbons of sand then accumulated in response to the action of waves and tides.

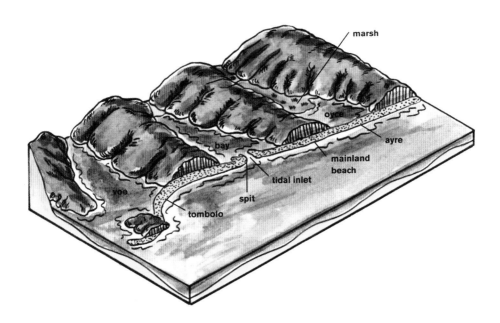

A tombolo is the name given to thin stretches of sand that link islands to the mainland. This is St Ninian's tombolo, which connects St Ninian's Isle to Mainland Shetland. During a ferocious storm in January 1993 when the *Braer* oil tanker ran aground causing untold environmental damage, this ribbon of sand completely disappeared as the wind and waves exerted a hugely destructive force. But over the years, in calmer conditions, the sand spit link to St Ninian's Isle has been restored by nature.

today as the voes of Shetland that are such prominent features of the contemporary landscape. Deposits of peat are to be found submerged to depths of 8 or 9 metres on the floor of some of Shetland's voes, illustrating the extent of sea-level rise since the end of the last glaciation.

The steeply sloping nature of the bedrock at the coast has hindered the formation of sandy beaches along many stretches of Shetland coastline. Instead thin ribbons of sand, known as ayres and spits, have formed. In some instances, they have blocked the end of the voes, and marshes have developed as a result.

A battered coastline

Once the land had emerged from its frozen mantle of snow and ice, the coastline began to take shape. The rocks forming the coastline were hard for the most part, built from ancient gneisses and schists, lavas, granites and sandstones. For thousands of years, the elements have gone to work on these rocks, battering them with Atlantic storms of great ferocity, with horizontal rain and howling winds. These elemental forces of nature have left their mark on the coastal rocks that define these islands.

One of Scotland's most iconic natural structures is the Old Man of Hoy in Orkney. The sandstone pillar stands on a plinth of volcanic lava and it has withstood the ravages of the elements for millennia.

The Old Man of Hoy is one of the most recognisable and spectacular sea stacks in Britain. It was formed by the sea exploiting weaknesses in the sandstone cliff and eroding it into the shape of an arch. Over time, the arch collapsed leaving the supporting pillar as an isolated stack. It will eventually collapse but there are many such structures in the making right now that will replace it. Headlands are currently being hollowed out to form arches, which will then collapse perhaps hundreds of years later.

This diagram illustrates how the sandstone cliffs are carved open by the elements. Sea stacks, caves, geos, blow holes and headlands are the landforms that result from this direct, violent and never-ending assault.

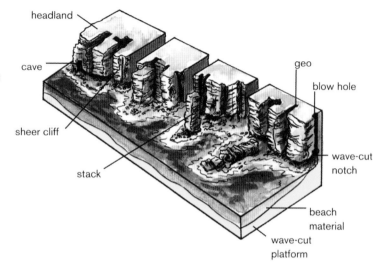

Dore Holm is an island close to Eshaness on the west of Mainland Shetland. The arch was created by the marine erosion. In time, the arch will collapse and a new sea stack, like the Old Man of Hoy, will be created.

Eshaness coast in western Shetland is made from tough stuff – mainly lavas – but wave action has cut them into a series of spectacular coastal landforms.

9
Tsunami!

Around 8,100 years ago, Orkney and Shetland were engulfed by a tsunami – a giant wave. An earthquake on the Norwegian continental shelf triggered a slide of massive quantities of rock, sand and gravel into deeper waters of the North Sea. This is known as the Storegga slide. This movement of material displaced huge volumes of sea water that subsequently caused fast-moving waves to radiate in all directions outwards from the site of the slide. When these waves encountered shallower waters, they reared up to 10m or more in height and broke onto dry land as a massively destructive force. The islands were uninhabited at this time, so there was no loss of human life here, although natural ecosystems would have been temporarily overwhelmed. The evidence for this catastrophic event is to be found in sedimentary layers at a number of locations across the islands. The layers that tell the story of this event are around 30cm thick and consist of sand, lumps of wood, pebbles and peat. The sands contain the remains of marine shells, including sea urchins and microscopic marine organisms. The presence of marine shells tens of metres above the present-day high-water mark is difficult to explain in any other way.

Orkney and Shetland felt the full force of the tsunami wave over 8,000 years ago.

The effects of the tsunami were widespread down the east coast of Britain, the west coast of Norway and the Faroe Islands. Similar tsunami-related deposits have been found at all of these locations. But Orkney and Shetland were in the front line of this ferocious assault by the sea.

10
Early occupation

The earliest human inhabitants of these islands settled in Orkney over 5,000 years ago, during the Neolithic or New Stone Age times. Evidence for this is provided at Skara Brae, which is regarded as the best-preserved collection of buildings and related artefacts of this age in Western Europe. This settlement predates Stonehenge and the construction of the pyramids in Egypt. Its discovery in 1850, after a massive storm, created great excitement among historians and archaeologists. It illuminated how people lived during those times and demonstrated how they worked, the tools they used and the food they ate.

The houses have no roofs today, but historians think that originally turf, seaweed or straw was used as protection from the elements. The furniture – beds, dressers, shelves and hearths – was sturdily constructed, hewn from the local sandstone.

The small population of hardy souls who lived here were fishermen, hunters and also farmers. We know that from the tools,

Remains of the houses at Skara Brae, Orkney.

Ring of Brodgar standing stones, Orkney.

seeds and bones found within or close to the site. They were livestock farmers who kept cattle and sheep, and grew crops in nearby fields to supplement their diet. These discoveries are significant as they relate to some of the earliest farmers in Britain. Before the discovery of this scene of domesticity, people from Neolithic times were thought to be hunter-gatherers with few farming skills.

No weapons have been found at this site, so historians assumed these early settlers were peaceful folk who had little need to defend themselves against attack. Instead, craft materials such as needles and buttons, ornaments and pottery have been found, so these early inhabitants were accomplished craftsmen and craftswomen.

The houses were abandoned around 4,500 years ago and it appeared as though the folk living here left quickly, abandoning many of their possessions. Speculation continues but the real reason for the inhabitants leaving this place in such an apparent hurry may never be known.

Standing stones are also an important part of the early story of the human occupation of Orkney. The Ring of Brodgar is perhaps the most impressive, originally consisting of 60 standing stones cut from the local sandstone. These pillars of rock create a circle that is around 100 metres in diameter. Archaeological investigations continue on Orkney, especially at the nearby site of the Standing Stones of Stenness.

Collectively, these remains, which have provided so many important insights into the life and times of the early inhabitants of Orkney, have been designated as a World Heritage Site.

11
The machair – a living landscape

The Atlantic-facing coast of the Outer Hebrides is internationally famed for its machair lands, but on Orkney and Shetland there are also fragments of this habitat to enjoy. Discrete pockets of machair are to be found from the tip of Unst down to Mainland Orkney. This coastal fringe is built from the debris of broken shells, transported onshore by the relentless winds and tides. This has created a globally rare natural habitat that is farmed by crofters, and has been for generations, in a low-intensity manner. Red clover, bird's-foot trefoil, orchids, yarrow and daisies grow in profusion during the season, and the machair is also an ideal habitat for birds, particularly lapwing, snipe and dunlin. The grasslands are alive with the sound of birdsong and, during the summer months, with a vibrant display of colour, courtesy of the rich variety of flowering plants.

Machair is a remarkable living landscape to be nurtured for the benefit of future generations.

12
Offshore oil and gas

The basis for the offshore oil and gas industry is an important part of Shetland's geological story. The infrastructure of pipes and storage

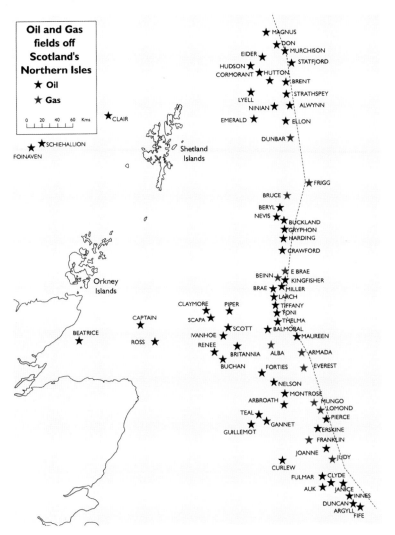

The search for oil and gas has led to the discovery of many sub-surface reservoirs of hydrocarbons that continue to yield great financial riches.

tanks that hold and distribute the oil and gas is clearly visible today at Sullom Voe and other locations across the islands, as are the many drilling rigs and supply vessels deployed offshore. But the sub-surface structures and geology that created this oil and gas bonanza are well hidden beneath the waves.

The oil and gas fields are located along a north–south axis, which is entirely related to the sub-surface geological structures. The main structure is known as the Viking Graben: a technical term for a trench on the sea floor that is defined by a fault system either side of the down-faulted block. This has helped to trap the oil in well-defined oil fields.

The source of the oil is rocks of Upper Jurassic age, the Kimmeridge Clay Formation. These rocks were laid down in the area now occupied by the North Sea around 155 million years ago. The waters at that time were rich in microscopic organisms, such as plankton, which accumulated on the sea bed after death. This organic material was subsequently covered by a build-up of layers of sands and mud and became deeply buried. Over time, the organic material

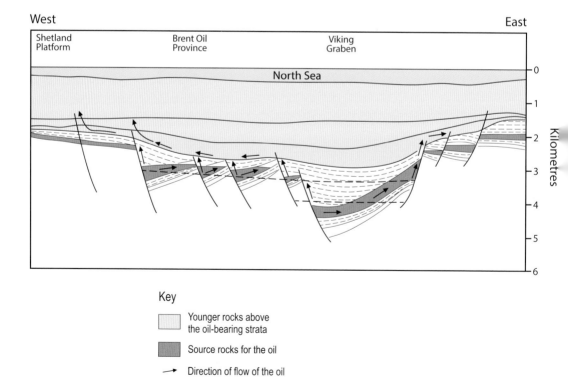

This is a vertical slice from the sea floor to the rocks beneath to demonstrate how oil and gas fields formed.

underwent a chemical change and matured to form oil and gas. The resulting hydrocarbons were less dense and were mobile within the layers of sediments, so they migrated upwards through the layers of younger strata until they met an obstruction or trap, such as a non-porous cap rock, where they built up in considerable quantities. The rock layers in which the oil and gas collect are known as reservoir rocks; they must be highly porous and permeable, so that oil and gas accumulations can be recovered and brought to the surface by drilling rigs.

Current estimates are that there are still around 17 billion barrels of oil or equivalent gas to be extracted from fields that have already been identified as viable enterprises in the waters around Scotland; these would keep activity going until 2050. If the price of oil rises and stays at a higher level, then the quantities are likely to increase as more oil fields become economically viable.

Sullom Voe on Shetland's Mainland has been the focus of oil and gas activity since 1975. It temporarily stores the hydrocarbons delivered by pipelines from the main offshore fields and subsequently feeds sea-going oil tankers that take the crude to refineries where useful products are manufactured. New fields have recently been discovered west of Shetland, so the commercial life of this facility has been extended as a result.

13
Places to visit

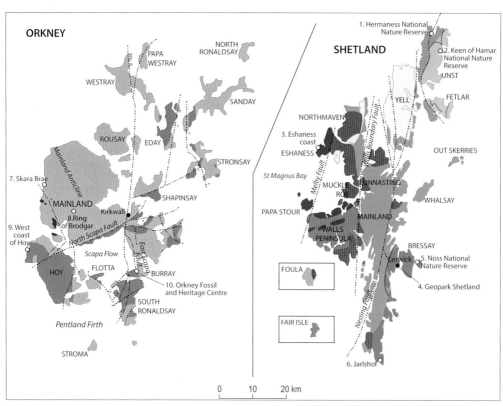

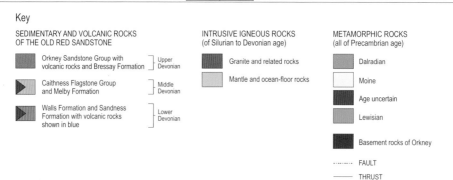

PLACES TO VISIT

Orkney and Shetland offer many possible places of natural interest to visit. What follows are just a few suggestions. Some are also of archaeological and wildlife interest. The OS Landranger (1:50,000 scale) maps will help you to navigate your way around the islands. The 'Bedrock Geology UK North' map, published by the British Geological Survey, will also be helpful in planning your visit.

1. **Hermaness National Nature Reserve, Shetland**: this is located at the most northerly point of Unst, Shetland, and has a geology comprising Dalradian schists. The reserve is primarily famed for its colonies of sea birds, including puffins, gannets, fulmars and guillemots. There is a visitor centre at the southern end of the reserve, and more information, including the reserve leaflet, can be found on the Scottish Natural Heritage (SNH) website.

Sea bird colony at Hermaness, Unst.

2. Keen of Hamar National Nature Reserve, Shetland: this reserve is located to the south of Harold's Wick on Unst. The geology of the site is described on pages 17 and 18. There is an access road leading from the A968 to a car park. From there, a track runs northwards into the reserve. There is more information, including the reserve leaflet, on the SNH website.

View across Keen of Hamar, Unst.

3. Eshaness coast, west Shetland: this coastline provides a wonderful scenic walk starting from the Eshaness lighthouse, travelling northwards. The dramatic cliffs are made from a thick succession of lavas and ash flows, indicative of a prolonged and violent volcanic episode.

4. Geopark Shetland: UNESCO Geopark status is given to areas of outstanding geological heritage value – in this case all of Shetland. Geoparks tell the geological story of their area through interpretative leaflets, panels and educational initiatives. To explore this Geopark, managed by the Shetland Amenity Trust, start at the Museum and Archives in Lerwick. Staff lead guided walks and deliver lectures to the public. More information about events can be found on the website.

The sandstone cliffs of Noss, Shetland, provide excellent exposure of the Old Red Sandstone and also provide a home for the sea bird population.

5. **Noss National Nature Reserve, Shetland**: this is located on an island off Bressay, Shetland, but it's worth the trip, particularly for birdwatchers. The geology comprises Old Red Sandstones, forming dramatic sea cliffs that are almost 200 metres high. Gannets, great skuas, puffins and guillemots are here in abundance. Access details are available on the SNH website.

6. **Jarlshof, Shetland**: this is the most remarkable archaeological site on Shetland. It has historical echoes from the Neolithic Period, Bronze Age and Iron Age, as well as buildings constructed during the Norse era. The site is in the care of Historic Environment Scotland.

7. **Skara Brae, Orkney**: this fascinating property is in the care of Historic Environment Scotland. It is signposted from Skaill on Mainland Orkney. There is adequate parking, good interpretative panels and leaflets, toilets and refreshments on site. It really is a 'must visit' destination.

Skara Brae, Orkney.

Ring of Brodgar, Orkney.

8. Ring of Brodgar, Orkney: these famous standing stones are located beside the B9055 road 8km north-east of Stromness, Orkney. There is adequate parking, but there are no visitor facilities.

9. West coast of Hoy, Orkney: this area provides a slice through the layers of sandstones that make up Orkney. There are no roads that access the west coast of the island, so it's boots on if you want to see these rocks, sea cliffs and, of course, the iconic Old Man of Hoy. The island has good accommodation and can be reached by car ferry, which runs from Houton (Mainland Orkney) to Lyness (Hoy), and by foot ferry from Stromness via Graemsay to Moaness (Hoy).

West coast of Hoy, Orkney.

10. Orkney Fossil and Heritage Centre, Burray: this heritage centre, south of Kirkwall, has interesting interpretive displays about the fossils and geology of Orkney and, possibly of equal importance, a tearoom!

Acknowledgements and picture credits

Thanks are due to Professor Stuart Monro OBE FRSE and Moira McKirdy MBE for their comment and suggestions on the various drafts of this book. I also thank Debs Warner, Mairi Sutherland, Andrew Simmons and Hugh Andrew from Birlinn Ltd for their support and direction. Mark Blackadder's book design is up to his usual high standard. Scottish Natural Heritage, in association with the British Geological Survey, published the *Landscape Fashioned by Geology* series that was the precursor to the new *Landscapes in Stone* titles. I thank them both for their permission to use some of the original artwork and photography in this book. I wrote the original text for *Orkney and Shetland – A Landscape Fashioned by Geology*. I have dedicated this work to Professor John Gordon. He was my colleague for over twenty years in Scottish Natural Heritage and before that with the Nature Conservancy Council. We worked on many projects together, including our first venture with Birlinn Ltd: *Land of Mountain and Flood*. His scholarly approach and attention to detail are amongst his many qualities. Together with a few others, we established SNH's earth science presence and function in Scotland in the early 1990s.

Picture credits

2–3 Bildagentur Zoonar GmbH/Shutterstock.com; 6 Richard Clarkson/Alamy Stock Photo; 10 © pixocreative.com; 12 (top) drawn by Jim Lewis, (lower) Richard Bonson/SNH; 13 Scottish Viewpoint/Alamy Stock Photo; 14 drawn by Jim Lewis; 15 (top) Allgord/Shutterstock.com, (lower) drawn by Jim Lewis; 16 drawn by Jim Lewis; 17 © pixocreative.com; 18 Krystyna Szulecka Photography/Alamy Stock Photo; 19 Prisma by Dukas Presseagentur GmbH/Alamy Stock Photo; 20 Zdenka Mlynarikova; 21 drawn by Jim Lewis; 22 (lower) © pixocreative.com; 23 (top) drawn by Robert Nelmes, (lower) Nigel Trewin; 24: (top) reproduced with permission from Nigel Trewin, *Scottish Fossils*, Dunedin Academic Press, 2013; 25 Niwat.koh; 26 (top) Lorne Gill/SNH, (lower) Adam Seward/Alamy Stock Photo; 28 (collage, from top right) SNH; Denis Burden/Shutterstock.com; fboudrias/Shutterstock.com; koal_a/Shutterstock.com; 29: Lorne Gill/SNH; 30 Ian Kirkwood; 31 (top) Lorne Gill/SNH; (lower) Craig Ellery/SNH; 32 Lorne Gill/SNH; 33 Chrispo/Shutterstock.com; 34 (top) Craig Ellery, (middle) Lorne Gill/SNH, (lower) Zdenka Mlynarikova/Shutterstock.com; 35 skapuka/Shutterstock.com; 36 Alan McKirdy; 37 john braid/Shutterstock.com; 38 Lorne Gill/SNH; 39 RCAHMS; 40 © pixocreative.com; 41 Sheila Halsall/Alamy Stock Photo; 42 © pixocreative.com; 43 Blue Gum Pictures/Alamy Stock Photo; 44 David Tipling Photo Library/Alamy Stock Photo; 45 Lorne Gill/SNH; 46 (top) Alan McKirdy, (lower) rphstock; 47 LatitudeStock/Alamy Stock Photo.